SERIES FOR CLI

Improving Clinical and
Professional Knowledge

Rehydration and Dementia: A Review of the Literature

Carole Archibald

May 2005

ISBN 1 85769 187 3

Contents

Acknowledgements

Firstly, I need to acknowledge the energy and drive of Frank Hitchman and the Dementia Services Development Trust who raised the money to enable this literature review to be undertaken.

Next a special thanks to Eileen Richardson, Library & Information Services Manager and Catriona Grant, Library & Information Assistant who helped in searches and obtaining articles, without whom my task would have been so much more difficult.

About the Author

Carole Archibald is a registered nurse and health visitor. She has worked in the field of dementia for the past 19 years and was awarded a PhD in 2002. She worked initially as a specialist health visitor then for 14 years as a Senior Fieldworker at the Dementia Services Development Centre (DSDC) at The University of Stirling. She is now an Associate Consultant and Trainer with the DSDC. She has worked extensively in the acute sector with staff, particularly nurses, to look at issues when people with dementia are admitted to an acute hospital. Over the past year she has undertaken a feasibility study with community health and social work staff in two areas of Scotland to look at whether interventions such as subcutaneous fluids and increasing staff awareness of dehydration in the community, will help to avoid some people with dementia being admitted to acute hospitals. This literature review and the forthcoming accompanying practice guide are outputs from the work on rehydration.

1 Introduction

The intended audience of this review is nurses and other practitioners who work with people with dementia in community settings and care homes. The aim is to provide an accessible yet systematic look at the literature so that practitioners can avail themselves of current research on the topic of dehydration and the interventions used to correct this condition. The premise of the review is that some acute hospital admissions of people with dementia may be avoided with interventions to correct either acute mild/moderate dehydration or chronic dehydration.

In this review I am seeking to explore, within the current literature, the use of subcutaneous fluids (SCFs) and other interventions, in the prevention or treatment of mild/moderate dehydration in people with dementia. The primary focus will be on the provision of SCFs, and I will use the term SCFs as opposed to hypodermoclysis, as this is the term that most practitioners use and are familiar with.

Dehydration is often a difficult clinical diagnosis to make as the clinical signs are either absent in older people or ambiguous with only laboratory evaluation providing a clear-cut diagnosis (Thomas et al 2004). There are differing opinions on how to diagnose and treat dehydration, and these will be explored. Most studies point to dehydration as the most common fluid electrolyte imbalance in older people (Hodgkinson et al 2003) and it is often the result of poor fluid intake during a febrile condition such as chest or urinary infection (Rochon et al 1997).

Some report that dehydration is present in as many as 25% of older people, particularly febrile residents in nursing homes where referral to an acute hospital is often necessary (Ferry et al 1999). Older people living in long-term care drink less and have less variety in what they drink, compared to those living in the community (Mentes 2000). It is not only those in institutional care that are at risk, however, older people living in the community also need to be considered (Bennet et al 2004).

An Italian study (Molaschi et al 1997: 156) looked at 2894 older people admitted to acute hospitals, and noted that under certain conditions, for example, respiratory tract infections, fluid loss primarily from the skin and respiratory tract (due to tachypnea) may increase three to five fold. In the absence of an increase in fluid intake, severe dehydration can result.

One in five older patients who are dehydrated because of febrile episodes are admitted to acute hospitals and around 7% of older patients admitted to hospitals in the US have a diagnosis of dehydration (Rochon et al 1997).

How to prevent, or address dehydration in older people, and therefore possibly avoid admission, is important, but becomes crucial when dementia is a factor. This is due to a number of reasons including increased mortality, morbidity and delays in discharge that occur in this patient group when they are admitted to acute hospitals compared to other older people who do not have dementia (Manning 2002). Cheek (2004), discussing older people and acute care, looks at the work of Andrews (2002: 223). From this she notes that different places or institutions have different values and operational codes. They do different kinds of work and are controlled by different kinds of knowledge and power. This is true of acute hospitals that have as a primary aim 'curing people' and an emphasis on the physical.

Cheek (2004), speaking about Australia but relevant in the UK, notes that older people are the core business of acute hospitals. Care of older people has long been seen by some as a Cinderella-type service, with the real work being undertaken in acute hospitals. The reality now is that due to demographic changes, older people are represented in most acute hospital areas. Acute hospitals have yet to adapt to this change of emphasis, particularly with regard to the admission of people with cognitive impairment. The curative focus in acute hospitals may detract from and not place such a high priority on developing appropriate attitudes, knowledge and specific skills needed for the care of older people (Cheek and Gibson 2003: 134).

People with dementia for example, often suffer psychological distress when they are admitted to such a frenetic environment as an acute medical ward. They tend not to cope well and may be further disabled in this kind of environment (Archibald 2002). The issues of people with dementia are discussed in more detail in other publications (Archibald 2003a, Archibald 2003b and Archibald 2002).

The review begins by describing the search strategies used to obtain the relevant articles, followed by a discussion of the quality and levels of evidence available. The research questions posed are noted and then each question is addressed in some detail. Whether there are safe options other than acute hospital admission when treating dehydration in people with dementia is discussed. The conclusions end the review.

2 Search Strategies

The search strategies included:

- Searches of appropriate electronic databases (for example, Cochrane Library, Medline, Cinhal, ASSIA, Age Info, British Nursing Index, and Psych Info).
- Reference checking of articles retrieved.
- Checking relevant internet sites.
- Hand searching of journals.

Key Search Terms

Dementia, Alzheimer's, older people/elderly: dehydration, hydration, Hyponatraemia/hypernatraemia, SCFs, hypodermoclysis, intravenous fluids, rapid response teams, fluid rescue, home support, hospital at home schemes, readmission, discharge.

3 Levels of Evidence - An Overview of the Literature

The evidence to support the use of SCFs in older people is mainly low level with a number of case studies reporting its useful application in older people, i.e. older people as opposed to people with dementia. The search terms for dementia specifically provided few studies.

To date there is only one systematic review that looks at the evidence for SCFs to treat dehydration in older people (Rochon et al 1997). In this study the main purpose is to evaluate the evidence supporting the use of SCFs to treat older people with dehydration and to discuss their use in long-term care settings.

Of those studies looked at in the systematic review, 18 met the inclusion criteria of the authors and were evaluated according to the type of fluid administered. The majority of studies reviewed were of poor quality. From the limited evidence available they suggest that this type of fluid administration can be safely used for electrolyte-containing fluids. Whether hyaluronidase is necessary to promote subcutaneous absorbtion remains unresolved. They note that SCFs have fallen out of favour because of adverse reactions related to previously used infusions of electrolyte-free or hypertonic solutions, types of fluid replacement that are unlikely to be used today (Rochon et al 1997).

Rochon et al's is the only systematic review with regard to SCFs, but was itself critised by the NHS Centre for Reviews and Dissemination (1997) for failing to include key features of the systematic review process (Moriarty and Hudson 2001). Since Rochon et al's study, there have been only two systematic reviews undertaken that address the research questions of my review and these are Hodgkinson et al (2003) and McGee et al (1999). The former looks at oral fluid intake and older people, the latter reviews the evidence for the physical diagnosis of hypovolaemia.

Hodginson et al (2003) look at older people and the focus is oral hydration. In this review the primary outcomes were those related to the identification of dehydration and the maintenance hydration or reversal of dehydration and so is relevant to this literature review. There were 17 articles that met the criteria for inclusion but again the majority of the studies were flawed in research design and most were observational or non-randomised, presenting little in the way of validated data. The authors note that, despite the high risk of dehydration in older people, there has been little quality research undertaken in this area, so their findings must be viewed with caution. This Australian study points to no clear determinants of the risk factors of dehydration, suggesting the need for fluid balance charts and testing the specific gravity of the patient's urine as the best methods of monitoring daily intake of fluids.

In a US study, McGee et al (1999) systematically reviewed the physical diagnosis of hypovolaemia in patients. Whilst this paper does not have as its focus SCFs it is useful in pointing to the difficulties in diagnosing dehydration in patients. Two types of studies were accessed. The first looked at the postural vital signs and capillary refill time of healthy volunteers included because they were the only studies that compare the physical signs with objective measurements of blood loss. The second included patients presenting at A & E with suspected dehydration due to vomiting and diarrhoea or reduced oral intake. It is the second that is of interest here.

In the latter studies, mainly older people were recruited (three studies: n=202, n=55, numbers not provided for third study). Here, it was not possible to show that any of the signs/symptoms were robust enough as indicators of dehydration, rather they used the presence of moist mucous membranes, the absence of sunken eyes and the absence of furrows on the tongue as signs that the person was sufficiently hydrated. For example, in one study they reviewed, the authors noted no correlation between dryness of mucous membranes and the degree of hypovolaemia. In two further studies, poor skin turgor proved to have no diagnostic value.

Slesak et al (2003) in a German prospective, randomised open clinical trial sought to compare intravenous versus subcutaneous infusions to treat dehydration in older people. This study was conducted on geriatric wards in Germany for a 20 month period where patients admitted with mild-to-moderate dehydration were randomly allocated to receive intravenous (IV) fluids or SCFs, with 48 in each group.

Again this small study is concerned with older people rather than dementia specifically, and the authors note that both types of rehydration procedures are equally well-accepted by these patients. In those who are confused however, they found that SCFs represent by far the superior method. These authors note also that since the 1980's, SC infusions have increasingly been discovered as an alternative to IV fluids particularly in the field of palliative care and care of older people, yet remains an underused procedure.

In the palliative care literature, Grande et al (1999), in a randomised controlled trial, looks at whether hospital at home facilitates death at home. What is evident is that SCFs are considered safe, easy to administer and that insertion of the cannula requires little training, with some suggesting that it can be undertaken by family members (Steiner 1998).

The only recent specific study that relates to hydration and mental health problems (Brown and Marland 2002) is a four-month investigation into patient hydration on one admission unit in Carlisle. This small study is about older people generally, and does not detail whether the 23 patients in the study had dementia or other mental health problems. Other papers obtained are small-scale and/or case study design or are descriptive. There are several papers that provide good practice ideas that will be of benefit in addressing at least one of the research questions and these include Sasson and Shvartzman (2001), Donnelly (1999), Abdulla and Keast (1997), and Noble Adams (1995).

Britton and Russell (2003), in a systematic review of the literature that considered multidisciplinary interventions for delirium in patients with chronic impairment, found no studies for inclusion. None of the studies looked at people who were known to have cognitive impairment prior to the onset of delirium and so could not be included. These authors point to the dearth of studies of delirium and chronic cognitive impairment yet this is precisely the group that are at high risk of delirium. Of the studies considered they found no reliable evidence for 'best management' of delirium other than the treatment of precipitating factors (Britton and Russel 2003) which includes dehydration.

In a US study, Fick et al (2002) have undertaken a systematic review looking at delirium superimposed on dementia that relates to research question 3 and so is relevant. Fourteen studies are reviewed, specifically those addressing delirium imposed on dementia. This systematic review of studies, mainly of hospitalised or nursing home patients, highlights the dearth of research in this area. Only one study systematically identified associated factors and interventions for delirium superimposed on dementia, but several were able to identify adverse events including accelerated long-term cognitive and functional decline and repeat admissions.

Of the studies reviewed, those given the highest study quality ratings found that those with delirium superimposed on dementia had lower mini-mental scores, and that delirium occurred more in those with late-onset dementia and those in the later stages, than in patients with other subtypes. One study (Sandberg et al 1999) pointed to 26% with hypoactive delirium and 22% with hyperactive. The fact that so many patients had hypoactive delirium is important as in these cases, the diagnosis of delirium may be missed.

One study (Lerner et al 1997), less highly rated 7 out of 16, showed delirium was associated with urinary tract infection, stress and surgery. Only one study, rated 9 out 16 in terms of quality rating (Mentes et al 2001), specifically pointed to inadequate fluid intake increasing the risk of dehydration.

The systematic review points to the high prevalence of delirium superimposed on dementia (22-89%) in both US community and hospital populations and the frequent lack of recognition of this condition by staff. The need for early recognition and prevention of delirium in people with dementia is stressed due to the high morbidity and mortality rates in this patient population when delirium occurs.

Within the limitations of the literature noted above, the research questions will now be addressed.

4 Research Questions to be Addressed in the Review

In order to explore the issues of prevention and treatment of mild/moderate fluid loss in people with dementia, several questions need to be addressed.

a) What is dehydration and why is it important to diagnose?

b) How safe are SCFs as a procedure to correct dehydration in people with dementia?

c) How is dehydration implicated when a person with dementia is admitted to an acute hospital?

d) How do health and social care staff work towards maintaining an adequate consumption of liquids in those people with dementia living in the community?

e) What are some of the difficulties involved?

f) Are there interventions they would recommend which are currently not available?

g) What are the anticipated advantages and disadvantages of these interventions?

h) What is the awareness of the importance of fluids in the prevention of illness in family carers and people with dementia?

4a) What is dehydration and why is it important to diagnose?

The clinical diagnosis of dehydration is often difficult to make yet the consequences of missed diagnosis are critical in terms of increased morbidity and mortality.

Thomas et al (2004), in a US study of 102 consecutive medical admissions of people 65 and over, found that as many as a third of physicians made an incorrect diagnosis. These authors note that a definitive diagnosis can only be made based on lab results such as when serum osmolarity is greater than 295 milliosmols and the BUN/creatine ratio was greater than 20. Bennet et al (2004) reviewing the literature suggests that the discrepancy of findings in various studies with regard to the number of people with dehydration admitted to hospital suggests that the ICD-9 definition of dehydration may underestimate the actual prevalence (Bennet et al 2004: 23).

Others, Hodgkinson et al (2003), argue that there is no absolute definition of dehydration, whilst Rafferty (2004) (Box 1) provides a working account. In dehydration, rapid weight loss greater than 3% of body weight due to loss of fluid or dehydration arising from one of two reasons, occurs. The first is water depletion the other is sodium depletion where there is a concomitant decrease in water (Hodgkinson et al 2003: 3).

Suhayda and Walton (2002) in a useful US practice-based paper, note that fluids and food are the major sources of water gain. Normal renal excretion is responsible for most water loss with the average daily urine output 1,500ml (Suhayda and Walton 2002). People with a higher proportion of body fat are more susceptible to fluid imbalance that causes dehydration. Also with age, there is a decline in total body water by 10%-15% attributed to reduced muscle mass and a reduction in the renal regulation of sodium and water balance (Suhayda and Walton 2002: 269). All these factors place older people at greater risk of dehydration whatever the setting.

Water loss of 1%-2% impairs cognitive and physical performance. Whilst all adults are at risk of dehydration, older people are more vulnerable due to metabolic changes that occur with ageing and those changes listed above. Symptoms can be insidious and factors such as chronic mental illness, medications that enhance fluid loss or self-imposed fluid restrictions to manage incontinence can all contribute to dehydration. Classic signs of dehydration may be missing in older people until the problem is well-advanced.

Bennet et al (2004), in a US retrospective review of medical records, records of 185 older people who visited a hospital emergency department, introduces the concept of chronic dehydration. Whilst acute dehydration may be experienced by any individual during illness such as, sickness and diarrhoea and blood loss, chronic dehydration is a fluid imbalance of longer duration but is still significant. Chronic dehydration can result in altered absorbtion of medications, delirium, and increased risk of death (Bennet et al 2004: 22)

The findings from Bennet et al's study becomes pertinent when looking at older frailer people living in the community, where chronic dehydration in older people living in the community is common. Suhayda and Walton (2002) argue that because older people living in the community may go unmonitored by family or professionals for a long period of time, they may be at risk of chronic dehydration.

Box 1 Dehydration - Common Types

Hyponatraemia dehydration - this is a result of excessive loss of fluid through for example:

Trauma/burns/sickness/diorrhoea/vomiting resulting in salt depletion

Hypernatraemic dehydration - water intake is reduced to less than that needed to maintain fluid balance.

The onset of this type of dehydration can go undetected as there is initially little change in blood pressure and plasma/sodium levels. Thirst may be present. (Rafferty 2004).

In a systematic review that looked at the physical diagnosis of hypovolaemia(1) in adults due to vomiting/diarrhoea or decreased fluid intake, few findings have any proven utility (McGee et al 1999). Rigorous conclusions about the role of physical examination in assessing the volume hydration status of patients with lowered intake of fluids are difficult to make because of the lack of studies. The most helpful findings of the evidence reviewed shows that if the patient has moist mucous membranes, absence of sunken eyes and absence of furrows on the tongue, they are not dehydrated. Following on from Thomas et al (2004), McGee et al suggests that clinicians should measure serum electrolytes, serum blood urea nitrogen, and creatinine levels when diagnostic certainty is required.

Box 2 Clinical signs of dehydration

- If the patient has moist mucous membranes, absence of sunken eyes and absence of furrows on the tongue they are not dehydrated.
- Dry oral membranes and upper body weakness and dry axilla supports the presence of dehydration (Thomas et al 2004).

Dehydration in patients due to lack of sufficient intake of fluids may indicate substandard care if due to neglect (George and Rockwood 2004). However, in one US study gathering data about dehydration, it can be difficult to ascertain whether dehydration was a consequence of neglect or due to rapid fluid loss (Mukamel 1997). Mukamel (1997), in this well-conducted US study examined the use of the patient review instrument to look at hydration. This instrument is used in a reimbursement reporting system used by New York State to assess quality of care based on five risk-adjustment outcomes.

Mukamel (1997) found that whilst factors such as deterioration in ADL rates and rates of physical restraint met all the validation criteria, dehydration did not. It was difficult to distinguish between cases due to rapid fluid loss that do not imply poor quailty of care, and cases due to inappropriate fluid intake that might. The author suggests improving the way dehydration is defined so that potentially, a better assessment of quality of care in preventing dehydration occurs.

Note (1) The term hypovolaemia - volume depletion, describes the loss of sodium from the extracellular space- intravascular and inerstitial fluid that occurs after gastrointestinal haemorrhage, vomiting, diarrhoea and diuresis. Dehydration refers to losses of intracellular water that ultimately lead to elevation in plasma sodium concentration. In most clinical studies investigators put the two disorders together. Hypovolaemia is used here to refer to both conditions (McGee et al 1999).

Box 3 4a) Main points

- It is difficult to diagnose dehdyration clinically.
- Few findings have any proven utility in clinically diagnosing dehydration. The most reliable are that if the patient has moist mucous membranes, absence of sunken eyes and absence of furrows on the tongue, they are not dehydrated.
- Clinicians should measure serum electrolytes, serum blood urea nitrogen and creatinine levels when diagnostic certainty is required.
- Dehydration occurs either because of water depletion or sodium depletion where there is a concomitant decrease in water.
- Older people are at greater risk of dehydration whatever the setting.
- Classic signs of dehydration may be missing in older people until the problem is well-advanced.
- The average daily urine output is around 1,500ml
- Water loss of 1%-2% impairs cognitive and physical performance.

- Chronic dehydration is a fluid imbalance of longer duration, but is still significant.
- Older people living in the community, especially those who are frail, are at high risk of dehydration as they may go unmonitored by family or professionals for a long period of time.
- Whilst clinical signs of dehydration may not be obvious, those caring for people with dementia/older people should be aware of the possibility of chronic, if not acute, dehydration in all settings.

4b) SCFs- are they a safe option?

Box 4 provides a useful description of this procedure. The general consensus in the literature is that this is a safe and easy hydration technique suitable for mild-to-moderately dehydrated adults, particularly older people (Sasson and Shvartzman 2001). The risks involved are minimal if fluids are administered according to indications and guidelines (see Rafferty 2004, Sasson and Shvartzman 2001).

The most common adverse effects such as localised oedema, can be treated by resiting the needle and simple massage. Pain can be related to the insertion of the needle into underlying muscle as opposed to subcutaneous tissue or to an increase in infusion rate but risks are minimal when the needle site is changed daily and care is taken with the rate of flow (Sasson and Shvartzman 2001: 1578).

Contraindications are that SCFs should not be administered rapidly and in large amounts. It is contraindicated in patients who may be at risk from pulmonary oedema (e.g. those with congestive cardiac failure). Pulmonary oedema is a rare adverse effect with, for example, only a 0.6% occurrence in 600 patients hydrated by SCFs in one study (Sasson and Shvartzman 2001).

The general view is that use of SCFs is a positive if generally underused procedure; useful not only to correct dehydration but even more effective in its prevention (Ferry et al 1999: 6). Donnelly (1999), reviewing the literature explored the benefits of SCFs, noting that in one study where 1,850 SC infusions were given to 270 older people there were few complications and that SCFs administration was much safer and less distressing to the patient than IV infusions.

Box 4 Defining SCFs

The subcutaneous infusion of fluid is a method of hydration where 1-2 litres of fluid per day can be infused through a small needle or plastic canula placed into subcutaneous tissue, for example, abdominal wall, back, thorax or upper thigh. This tissue is mostly fat and the blood vessels in the fat quickly take up the fluid (AFP 2001). This procedure can be used for older people with mild, or at risk of mild dehydration (Rafferty 2004).

First described in 1913, SCFs have been used for years to provide fluid for those with inadequate oral fluid intake (Rochon et al 1997).

SCFs are indicated for those people who can no longer take adequate oral fluids, who are mild-to-moderately dehydrated and for whom it is difficult or impractical to insert an IV line. The main areas of use have been in palliative care or in the care of older people (Sasson and Shvartzman 2001).

Box 5 4b Main Points

- This is a safe and easy hydration technique suitable for mild-to-moderately dehydrated adults.
- The risks involved are minimal if fluids are administered according to indications and guidelines.
- Contraindications are that SCFs should not be administered rapidly and in large amounts. It is contraindicated in patients who may be at risk from pulmonary oedema.
- It is an underused procedure.
- It is less distressing to older people/people with dementia than IV infusions.

4c) How is dehydration implicated when a person with dementia is admitted to an acute hospital?

The studies reviewed are mainly from North America so care needs to be taken when applying the findings. Readers also need to be mindful about the earlier discussion of the difficulties in diagnosing dehydration in older people and the differences in how dehydration is defined.

Rochon et al (1997) in their review noted that 7% of older patients admitted to acute hospital suffer from dehydration. Copeman (2000) places the number higher at 10% and that as many as 25% of immobile patients suffer from mild chronic dehydration. There are a number of factors that increase the risk of dehydration including: reduced thirst response, swallowing problems, fear of incontinence, communication problems, depression and dementia. Copeman also suggests that when the older person relies on others to provide drinks, factors such as lack of variety, drinks not placed within reach or served in the wrong type of cup come into play and can impact negatively. People with dementia may have the capacity to eat and take fluids but may forget the steps needed to eat and drink (Amella 2004).

Box 6 Dehydration-Risk factors (Copeman 2000)

- Reduced thirst response.
- Swallowing problems.
- Fear of incontinence.
- Poor communication.
- Depression and dementia.
- Physical limitations including restraints

Other factors

- Lack of variety of fluids offered.
- Placing fluids out of reach.
- Wrong type of cup offered.

Molaschi et al (1997) also notes the increased prevalence rates of dehydration increased with age (85 years and above) with higher mortality rates evident. Their data points to a relevant association between advanced age and the onset of dehydration before admission to hospital. In these patients, often affected with such conditions as stroke, dementia and immobility, thirst sensation is impaired as is their ability to obtain water. In many of these cases dehydration is related to a febrile disorder.

Reviewing studies of nursing home residents admitted to hospital with dehydration, Hodgkinson et al (2003) found that in one study 6.5 per 1000 admissions were recorded in one hospital and was as high as 22.5 per 1000 admissions in another hospital. In another study mortality rates of older patients admitted with dehydration were as a high as 45-46%. These studies reviewed by the authors were undertaken in the 1980s but the findings remain pertinent twenty years later (Hodgkinson et al 2003).

Primarily in the literature, delirium is noted as one of the main reasons or complicating factors when older people and particularly people with dementia are admitted to hospital as a medical emergency (Manning 2002, Archibald 2002) with higher mortality rates, increased risks of institutionalisation and delays in discharge compared to those who do not develop delirium (Britton and Russell 2003, Fick et al 2002, Lyketsos et al 2000, George et al 1997). Britton and Russell (2003) note that 45% of those who are cognitively-impaired go on to develop delirium and these patients have longer stays in hospital, complications and poor outcome (Britton and Russell 2003). To reiterate, of the studies considered by these authors, no reliable evidence for 'best management' of delirium other than the treatment of precipitating factors (including dehydration) was found (Britton and Russel 2003).

Zeleznik's (2001) review of two systematic reviews on risk factors and effective preventative measures of delirium included 1,365 subjects with either incident or prevalent delirium and 5,513 controls. There were methodological concerns about the studies reviewed including the under-reporting of non-significant risk factors but there were some points of interest highlighted. Dehydration was not mentioned specifically but pertinent to this review, dementia was positively associated with delirium in 12 out of 15 studies as the strongest risk factor of a patient developing this condition. An association was also noted between abnormal serum sodium concentrations in four out of five studies.

In a paper that looks at a measure of the severity of delirium, the Delirium Index appears to be a reliable, valid and responsive measure of the severity of delirium in patients with delirium who have dementia or not (McCusker et al 2004: 1747).

In a study (Tolson et al 1999) that explored the components of best nursing practice in the care of those with dementia admitted to hospital (n=213 >65 years admitted to two acute medical wards) 66% of patients were found to have cognitive impairment with 40% of these experiencing severe problems. No evidence of formal assessment was located and only 15% of patients had anything in their care plans associated with chronic confusion.

This has implications when considering the risk factors associated with dehydration identified by Copeman (2000), that is dementia, poor communication skills and reduced thirst response.

In a guest editorial paper George and Rockwood (2004) note that dehydration is a predictor of impaired cognitive status. Archibald (2002 and 2003a) provides information for nurses about issues such as cognitive assessment and assessment of delirium.

Box 7 4c) Main Points

- Percentage of older people admitted with dehydration varies between 7% and 25%
- There are a range of risk factors that place patients at risk of dehydration including lack of thirst response (see Box 6).
- Prevalence rates of dehydration increase with age (85 years and above) with higher mortality rates evident in these groups.
- Of those who are cognitively-impaired, approximately 45% may go on to develop delirium and these patients have longer stays in hospital.
- Dementia was positively associated with delirium as the strongest risk factor of patients developing this condition.
- Studies in the prevention of delirium have concentrated on hospital rather than the community.
- Formal assessment of cognitive impairment is often not undertaken on admission despite dementia being a strong risk factor in developing delirium.

4d) How do health and social care staff work towards maintaining an adequate consumption of liquids in people with dementia living in the community?

Kayley et al (1996) is an unusual study, in that it looks at the use of IV antibiotic therapy in the person's own home. The patient group however are people with HIV infection, not those with a dementing illness, but the study provides some useful pointers. They treated 67 patients over two years that resulted in a saving of 2275 hospital days. In the study the existing network of NHS community nurses were trained by community specialist nurses in the safe administration of antibiotics, care of IV access lines and recognition and treatment of anaphylaxis.

The authors note that essential to the success of the programme was a multidisciplinary team working between hospital and the community that had a written shared-care protocol. Also the support provided by the hospital was another reason for its success - all patients, their GPs and community nurses had 24-hour access to the Infectious diseases unit of the hospital.

The above interventions introduced were more difficult than those using SCFs yet with certain provisos it proved to be possible to instigate. The authors conclude that this innovation will be of benefit to both patients and budgets (Kayley 1996).

Although some authors point to SCFs being easy and safe to use in any environment, including the patient's home (Ferry et al 1999), there are few studies that look at the provision of SCFs in such an environment. The majority of papers look at hydration in hospital or nursing homes. Most of the studies used in this section will focus on hydration in long stay settings.

The maintenance of adequate hydration for older people and particularly those with dementia is an ongoing challenge for staff/carers working in different environments, particularly nursing homes.

In long-stay care, all types of dehydration are encountered but the most prevalent type appears to be a subclinical state of dehydration or chronic dehydration due to the older person not drinking enough for whatever reason (Mentes and Culp 2003). Reviewing the literature, these authors point to a number of conditions that may result in both admission and repeated admission for the resident. These include acute confusion, dehydration, urinary and respiratory infections and constipation, often the consequence of chronic underhydration (Mentes and Culp 2003: 212).

The state of chronic underhydration means that the person is more susceptible to small environmental and physiological stressors that can precipitate dehydration and associated health problems (Mentes and Culp 2003). Mentes (2000) makes the point that often when an older person is admitted with, for example, a chest infection, the underlying antecedent condition of underhydration/dehydration is obscured and therefore may not be coded.

In their US quasi-experimental treatment and control group design with 49 participants, including those with dementia, from four nursing homes, Mentes and Culp (2003) tested the effectiveness of an eight week hydration intervention. Their study looked at whether hydration management would prevent or reduce urinary/chest infections and acute confusion in these resident populations. The intervention was based on the individual's fluid goal and adjusted for his/her weight. Strategies for ensuring adequate fluids included 180 mls of fluid intake with each medicine administered, fluid rounds morning and evening and 'happy hours' or 'tea time' twice a week in the late afternoon. Weekly urinalysis, including assessment of urine colour, and a 24-hour fluid record were undertaken to look at whether the person achieved adequate hydration.

The authors acknowledge the limitations of their study that include the difficulties with regard to sample size and randomisation issues but allowing for these and other methodological difficulties, their findings are interesting. Despite being more cognitively-impaired, more physically-impaired and more at risk from acute confusion, the treatment group had fewer hydration-related events (HLE's). This finding would resonate with those of Suhyada and Walton 2002) whose study is discussed later.

An important finding was that the interventions may have proven more efficacious if they were tailored to a person's specific risk factors of dehydration. They noted that individuals should be thoroughly screened for hydration problems on admission to nursing homes and hydration management strategies should be tailored to the individual who is at risk. These strategies should also be tailored to the time constraints of the staff, that is, they should be realistic given the often poor staffing levels and low levels of trained staff. A point made was that those with dementia need to be monitored closely. Often they are physically able but they may forget to drink if not prompted due to their cognitive impairment. Those with a fear of incontinence also need to be considered (Mentes and Culp 2003: 223).

In a controlled US clinical intervention trial, Simmons et al (2001) evaluated the effects of behavioural interventions to increase fluid intake in nursing home residents. In the first group, residents were given 16 weeks of four verbal prompts to drink each day in between meals. In the second group eight weeks of verbal prompts per day in between meals were given, and in the third group, eight weeks of verbal prompts per day in between meals plus compliance with participant drink preferences occurred. This study was conducted in conjunction with another that sought to improve incontinence and mobility in nursing home residents.

The conclusions drawn were that residents' response to verbal cues was significantly influenced by their cognitive status. Those with greater cognitive impairment showed the largest increases in their fluid intake in response to verbal prompts alone, whereas those less cognitively-impaired responded better when a choice of drinks was added to the prompt. This finding points to drink preference being an important component of obtaining resident compliance to increase fluids, particularly in those less cognitively-impaired. Another important finding was that, despite the increase of fluid intake due to the interventions, the fluid intake of the majority of participants was still inadequate so rendering them at risk of dehydration (Simmons et al 2001: 932).

Worth mentioning is that the common practice in long-term care settings of thickening fluids may contribute to dehydration on several levels. The substances for thickening are often unpalatable and refused. In one study that looked at the benefits of monitoring fluid thickness in the dietary management of dysphagic stroke patients, the authors suggested that such thickeners may lead to a reduced fluid intake (Goulding and Bakheit 2000).

Whilst not discussing community interventions specifically, Suhyada and Walton (2002) argue for strategies for maintaining hydration with different groups. These groups of staff would include the health team, the patient and the caregivers in order that a collaborative overall plan can be compiled. Education is a key component to alert staff and carers not only to maintaining adequate hydration, but in providing information regarding medications that affect the hydration status of the patient. These authors look at a wide range of reasons for dehydration including diarrhoea and vomiting and excessive exercise and suggest that as oral fluids are the easiest and least expensive these should be tried first, providing the patient is able to swallow and tolerate oral fluids.

Brown and Marland (2000) provide some ideas for nurses when assessing patients for dehydration, albeit in a hospital setting (Box 8).

Box 8 Assessing for Dehydration

- Blood samples analysed for urea and electrolyte levels.
- Mucous membranes observed for dryness.
- Urine samples tested for specific gravity.
- Patients questioned about fluid intake.
- Skin turgor tested.
- All medications assessed to look at start date, adherence efficacy and requirements.
- Patients were asked if they were constipated.

Brown and Marland (2000)

These ideas are useful but some of these questions might be difficult for people with dementia to answer. Also as McGee et al (1999) noted, skin turgor, proved to be of little diagnostic value.

Mental and physical impairment can affect functional ability and impact on the person's ability to recognise the needs for fluids or access the fluids needed. For example, a person with dementia may have a visual agnosia (Archibald 2002) or a person with arthritis or following a stroke may not be able to grasp and hold a cup or glass (Suhyada and Walton 2002). Any interventions therefore need to be mindful of these disabilities. Box 9 provides some examples of the initiatives that could be used to prevent dehydration in those with mental health problems, including those with dementia. Although these ideas were carried out in a hospital setting, most could be easily transferred to other settings.

Box 9 Prevention of dehydration

- A water tower with cooler installed.
- The kitchen left open at all times and patients encouraged to make themselves drinks.
- A selection of drinks made available to give patients a variety of choice.
- Visitors encouraged to use these facilities to make themselves and patients a drink.
- Jugs of juice placed on dining room tables at meal times and topped up regularly.
- Regular drinks rounds on the wards

(Brown and Marland 2000).

Others (Robinson and Rosher 2002) again point to the need for an individual plan of care but also suggest a colourful trolley, one yellow and one purple. These colours were selected because older people see these colours best. The trolleys were decorated according to the season/ holidays. Colourful jugs and glasses were bought so that made the whole thing look attractive to residents. There was a choice of four different kinds of cold drinks including juices, milk shakes, lemonade and a variety of hot drinks such as tea, coffee and hot chocolate. Their hydration programme set out to provide an additional eight ounce drink mid-morning and mid-afternoon.

In this small US study that included 51 participants in one nursing home, it was demonstrated that two additional glasses of fluid per day can raise total body water to a normal level for most of their residents. There were fewer falls, fewer urinary infections and fewer laxatives needing to be prescribed. Importantly they found that it was feasible for nursing assistants to devote five minutes per shift to providing additional fluids without this impacting too much on other duties. In order for this intervention to be successful, staff had to be aware of the rationale and the benefits that accrued not only to residents but to them as staff (Robinson and Rosher 2002 : 211).

Assessment of the person's risk factors (see Box 10) are important. Whilst it can be supposed that the more cognitively-impaired the person the greater the risks, one study found otherwise. In this study the findings pointed to significantly more dependent residents and cognitively-impaired residents having higher fluid intakes as these were deemed to be more identifiable by staff as requiring assistance. Paradoxically, the most at risk were the semi-independent residents who appeared capable enough to obtain fluids when they wanted them but who for a number of reasons, were actually not doing so (Suhyada and Walton 2002: 11). They also noted that a significant number of nursing home residents obtained fluids when receiving medication, so if residents did not receive fluids from this source, they would fall short of their daily requirement.

Observations of the signs of dehydration are crucial but as Suhyada and Walton (2002) note the signs can be subtle, deceptive or non-existent in older people. Since the thirst perception may be reduced, regular encouragement of fluids (daily 1,500 mls unless contraindicated) and noting down oral intake is important, particularly in those who have dementia. For those needing substantial fluid replacement, IV or SCFs may be needed. Any person at risk of volume overload will need to be monitored carefully during fluids replacement.

A simple oral fluid intake sheet and /or specific gravity test might be the best way of monitoring daily fluid intake with regular presented fluids to bedridden and semi-independent residents to maintain adequate fluid intake (Suhyada and Walton 2002).

Box 10 Risk factors associated with higher risk of dehydration

- Age
- Mobility
- Functional ability
- Gender (women more at risk?)
- Visual impairment
- Speaking ability
- Incontinence
- The number of opportunities for taking fluids

Hodgkinson et al (2003)

Other risk factors include the taking of multiple medications such as diuretics and laxatives (Robinson and Rosher (2002).

In a US retrospective chart review of 36 nursing home patients with a mean age of 85 years with significant cognitive impairment, SCFs were started prophylactically. A control group was not utilised. SCFs were initiated when nursing observation of inadequate oral fluids were noted and before obtaining laboratory results. They were discontinued upon resumption of adequate oral fluid intake (Hussain and Warshaw 1996: 970). In those that were given SCFs, extensive assistance or total dependence was noted with 86% having moderate-to-severe cognitive impairment. The most common indication for the use of SCFs was pneumonia (59%). Antibiotics were given orally or intramuscularly and SCFs given.

The findings from the study showed that the majority of patients improved clinically although significant improvement in laboratory blood tests were not apparent.

4e) What are some of the difficulties involved in the administration of SCFs?

There appears to be a general consensus in the literature that SCFs are safe if certain caveats and ground-rules are observed. What appears to be lacking in the literature is an imaginative leap with regard to the use of SCFs in areas other than nursing home or hospital environments such as in the person's own home from where the majority of admissions to hospital occur. There is a lack of any studies on SCFs and people with dementia in their own home. The only studies in the person's own home focus on palliative care (Moriarty and Hudson 2001). Here, one focus of the discussion centres around the ethical considerations of giving SC/IV fluids which might prolong life unnecessarily as opposed to addressing symptom control.

When considering the use of SCFs some attention needs to be given to staffing issues such as staffing levels and training. However the literature appears to suggest that once training and education of staff occurs and they are given a rationale for the procedure, the benefits become apparent. Many nursing home staff find the administration of IV fluids difficult whereas SCFs may be more acceptable to them (Hussain and Warshaw 1996).

There are some difficulties noted when SCFs are administered but these appear to be minor. For example, minor skin irritation and large bruises have been reported in patients with blood coagulation disorders and there can be localised pain (Hussain and Warshaw 1996).

4f) Are there interventions they would recommend which are currently not available?

Throughout the literature discussed above, in papers mainly from the US, SCFs are reported to be safe, simple to use and cost-effective in less acute care settings such as nursing/residential homes. However, although a number of patient groups could benefit from this procedure, including people with dementia, Moriarty and Hudson reviewing the literature (2001) note that to date, there is no literature that describes its use outside residential or acute care setting, that is in the person's own home. These authors have palliative care as their focus but acknowledging the benefits of SCFs for a wider patient group, set up a working party to develop guidelines for community nurses involved in providing SCFs in the patient's home. This paper provides useful pointers for any future pilot studies.

The authors look at the ethical issues, for example, in advocating artificial hydration in dying patients, a controversial area (Mion and O'Connell 2003, Fainsinger and Bruera 1997). There may be scope for some of these arguments to be aired when considering treating people with severe dementia who have developed a chest infection. The paper also takes into account family and client consent and education and discusses the purpose of SCFs and such factors as what will happen if the infusion is stopped. The provision of fluids in this study is to alleviate distressing symptoms, not to prolong life.

Following on from Mion and O'Connell (2003), Ersek's (2003) is also a useful paper looking at the clinical, legal and ethical implications in providing, or not, artificial fluids in palliative care situations. Bosek et al (2003) look at the factors that promote a good death for the person with Alzheimer's disease in nursing homes from a family carers' perspective. Using the standards suggested by Thomas and Day (2000) one of the standards is that death should occur in familiar surroundings. It is possible that providing SCFs when someone with dementia becomes acutely ill, may obviate the need for hospital admission and allow them to die in the community and thus meet this particular standard.

4g) What are the anticipated advantages and disadvantages of these interventions?

There are many disadvantages for people with dementia if they have to be admitted to an acute hospital as discussed previously. Several studies have pointed to a regime where extra fluids are provided several times a day outwith mealtimes, reducing the number of chest and urinary infections and also constipation in residents in nursing homes (Mentes and Culp 2003) where often many residents are chronically dehydrated.

Many nursing home staff find the administration of IV fluids difficult whereas SCFs may be more acceptable. A relatively safe procedure SCFs, if set up prophylactically have the potential to reduce 'costs' and the need to transfer people to hospital from nursing homes (Hussain and Warshaw 1996). These authors note that frequently treatment of acute illness or chronic unstable illness with the need for IV fluids was not attempted in the two nursing homes under study because of the lack of skilled personnel.

Staff had the capability to administer IV fluids but this procedure was rarely used, whereas SCFs as a method of hydration had been used in the homes for two years, treating this group of people in the nursing homes where staff were familiar was less traumatic and more cost-effective (Hussain and Warshaw 1996: 970). Also staff found the use of SCFs to be very convenient as it reduced set up time, monitoring and need for restraint compared to IV therapy (Hussain and Warshaw 1996: 972). See Box 11 for advantages and disadvantages of the use of SCFs).

Box 11 Advantages of SCFs (compared to IV fluids)

- Safe and easy to use
- Limited trained personnel needed
- Less monitoring needed
- Easy to reinsert
- Phlebitis, cellulitis infection, thrombosis are not usual problems
- Less pain and discomfort to the patient
- Less immobilisation required
- Less chance of fluid overload
- More acceptable to patient and staff
- Can be started and stopped without any problems
- Cost-effective

(Hussain and Warshaw 1996: 275)

4h) What is the awareness of relatives and people with dementia about the importance of fluids in the prevention of illness?

In the literature generally, family carers tend not to be discussed except in the passing as the focus of the papers reviewed is that of patients in institutional or acute care settings. In the grey literature the emphasis is mainly on food, but some publications discuss the importance of fluids in terms of wellbeing and prevention of illness (Copeman 1999).

Accessing the grey literature, it is apparent that many of the publications are aimed at professional carers. The Alzheimer's Society's (2000) Fact sheet on eating, Copeman (1999) Nutritional care for older people: a guide to good practice plus the Dementia Services Development Centre's (Stirling) Oh Good Lunch is Coming video and the Stickerpack (Crowe 2003) are all useful but tend mainly to provide information for professional carers. The Alzheimer's Society provides a very wide range of leaflets including a number on food but none specifically on fluids. So there appears to be a gap in both the grey and other literature on this topic for family carers and people with dementia.

5 Conclusions

Dehydration is the most common fluid electrolyte imbalance in older people, yet the lack of robust studies on this aspect of care is of concern. The quality of evidence is generally low level despite the general consensus that dehydration is an important factor in terms of morbidity and mortality in older people, particularly people with dementia, admitted to acute hospitals.

An important finding and one that resonates with a person-centred approach to care, is that interventions may be more effective if they are tailored to a person's specific risk factors of dehydration and knowledge of the person's cognitive status. For example, a resident's response to verbal cues may be significantly influenced by their cognitive abilities. Those with greater cognitive impairment in one study showed the largest increases in their fluid intake in response to verbal prompts alone. In those less cognitively-impaired, response was better when a choice of drinks was added to the prompt. These findings are easily applied in practice situations and so can be tested further by practitioners.

Increase of fluid intake is necessary but in one study, the fluid intake of the majority of participants was still inadequate after the research interventions of increased fluid intake. People with dementia and older people generally may be chronically dehydrated in whatever environment. Rather than waiting until acute dehydration occurs, more emphasis needs to be placed on a prophylactic approach.

Something as simple as two additional glasses of fluid per day may make a difference in terms of fewer falls, fewer urinary infections and fewer laxatives needing to be prescribed. Sufficient time and staff are always cited as difficulties, but one study showed that it is possible to provide even five minutes in a shift in nursing homes/residential homes to give additional fluids without this impacting too much on other duties. In order for this intervention to be successful staff have to be aware of the rationale and the benefits that accrued not only to residents but also to staff.

The signs and symptoms of dehydration in older people are more subtle and often not apparent until more pronounced. Therefore reliance on such signifiers as skin turgor or fissured tongue might not be so accurate. Risk factors such as poor fluid intake and medications need to be taken account of, but a seemingly easy way to guide staff is the colour and specific gravity of the person's urine. Education is a key component to alert staff and carers not only to maintaining adequate hydration, but in providing information regarding medications that affect the hydration status of the patient. It is important to stress these findings in staff education/ training materials.

The need for each older person in nursing homes/hospital to be cognitively assessed and also assessed for the risk of dehydration is noted but so too is the often low staffing levels and levels of trained staff available in institutional settings. The implication of this is that interventions, whether assessing for risk or providing fluids in whatever form, need to be easy and straightforward. Staff also need to be aware and understand the rationale for interventions for these to be successful.

Imaginative ways of providing fluids such as bright colourful cups and trolleys, and providing a choice of drinks appears a useful tactic. Also the awareness that groups of older people such as those who are concerned about incontinence or those who are able to help themselves to drinks but do not for whatever reasons, need to be monitored as do those with a diagnosis of dementia who may need to be prompted to drink. Staff need also to be aware of visual agnosia and other disabilities such as communication and behavioural disturbance in people with dementia.

SCFs appear to be a safe intervention if certain caveats are observed, are easy for staff to use, and are acceptable to older people with one study pointing their efficacy in the prevention of dehydration as well as being used to treat this condition. A lack of knowledge and therefore confidence in using what appears to be a safe and effective way of replacing fluids in those who are unable to take sufficient oral fluids means that some people, particularly those with dementia, may be admitted to hospital unnecessarily. SCFs appear to be used rarely but when they are used they are effective both in treatment of mild/moderate dehydration and importantly, in the prevention of dehydration when a person is unable for whatever reason to drink sufficient fluids.

Often when an older person is admitted to hospital with, for example a chest infection, the underlying antecedent condition of under-hydration/ dehydration is obscured and therefore may not be coded. The need for dehydration to be coded so that the extent of the problem can be elicited, is important when considering further research.

The research reviewed has been mainly North American. There appears to be a dearth of UK research undertaken in this area of work. Also there are gaps in research with regard to people with dementia and dehydration, and specifically, research into the use of SCFs in the person's own home. More research is needed in these areas as is more education of staff and families as to the importance of fluid intake in people with dementia.

References

Abdulla A and Keast J (1997) Hypodermoclysis as a means of hydration *Nursing Times* 93(29): 54-55

Amella EJ (2004) Feeding and hydration issues for older adults with dementia *Nurs Clin N Am* 39: 607-623

American Family Physician Hypodermoclysis : A way to replace lost fluids http://www.aafp.org/afp/2001 :1101

Archibald C (2002) *People with dementia in acute settings* Stirling: Dementia Services Development Centre

Archibald C (2003a) *People with dementia in acute hospitals: A practice guide for registered nurses* Stirling: Dementia Services Development Centre

Archibald C (2003b) *People with dementia in acute hospitals: A practice guide for clinical support workers* Stirling: Dementia Services Development Centre

Bennet JA, Thomas V and Reigel B (2004) Unrecognised chronic dehydration in older adults *Journal of Gerontological Nursing* 30(11): 22-28

Bosek MS, Lindeman DA, Burck JR and Gwyther LP (2003) Promoting a good death for persons with dementia in nursing facilities: familiy caregivers' perspectives *JONA's Healthcare Law, Ethics and Regulation* 5(2): 34-41

Britton A and Russell R (2003) Multidisciplinary team interventions for delirium in patients with chronic cognitive impairment *The Cohcrane Library* Issue 4: 1-11

Brown J and Marland G (2002) Hydration in older people with mental health problems *Nursing Times 98* (3): 38-9

Cheek J (2004) Older people and acute care: A matter of place? *Crisis and Loss* 12(1): 52-62

Cheek J and Gibson T (2003) Issues impacting on registered nurses providing care to older people in an acute care setting *Nursing Times Research* 8(2): 134- 49

Copeman J(2000) Promoting nutrition in older people in nursing and residential homes *Community Nursing* 5(6): www.dotpharmacy.com/upnusteld.pdf

Crowe V (2003) *Stickerpack: safe eating and swallowing* Stirling: Dementia Services Development Centre

Dementia Services Development Centre (2002) *Oh Good Lunch is Coming* (video) Stirling: Dementia Services Development Centre

Donnelly M (1999) The benefits of hypodermoclysis *Nursing Standard* 13(52): 44-5

Ersek M (2003) Artificial Nutrition and Hydration *Journal of Hospice and Palliative Nursing* 5(4): 221-30

Fainsinger RL and Bruera E (1997) When to treat dehydration in a terminally ill patient? *Supportive Care in Cancer* 5(3): 205-11

Ferry M, Dardinaine V and Constans T (1999) Subcutaneous infusion or hypodermoclysis: a practcial approach *JAGS* 47(1): 9 -5

Fick DM, Agnostini JV and Inouye SK (2002) Delirium superimposed on dementia: A systematic review *JAGS* 50: 1723-32

George J and Rockwood K (2004) Dehydration and delirium-not a simple relationship Guest editorial *The Journal of Gerontology* 59A(8): 811-112

George J, Bleasdale S and Singleton SJ (1997) Causes and prognosis of delirium in elderly patients admitted to a district general hospital *Age and Ageing* 26: 423-27

Goulding R and Bakheit AM (2000) Evalaution of the benefits of monitoring food thickeness in the dietary management of dysphasic stroke patients *Clin Rehabil* 14: 119-24

Grande GE, Todd CJ, Barclay SIG and Farquar MC (1999) Does hospital at home for palliative care facilitate death at home? Randomised controlled trial *BMJ* 319(7223): 1472-75

Hodgkinson B, Evans D and Wood J (2003) Maintaining oral hydration in older adults: a ststematic review International *Journal of Nursing Practice* 9(3): S19-28

Hussain NA and Warshaw G (1996) Utility of clysis for hydration in a nuring home residents *JAGS* 44(8): 969-73

Kayley J et al (1996) Safe intravenous antibiotic therapy at home: experience of a UK based programme *Journal of Antimicrobial Chemotherapy* 37(5): 1023-29

Lerner AJ, Hedera P and Koss E (1997) Delirium in Alzheimer's patients *Alzheimer's Disease and Associated Disorders* 11: 16-20

Lyketsos CG, Sheppard JM and Rabbins P (2000) Dementia in elderly persons in a general hospital *American Journal of Psychiatry* 157: 707-7

Manning W (2002) *Delirium* Stirling: Dementia Services Development Centre

McCusker J, Cole MG, Dendukuri N and Belzile E(2004) The delirum index, a measure of the severity of delirium: new findings on reliability, validity and responsiveness *JAGS* 52 (10): 1744 - 49

McGee S, Abernethy WB, and Simel DL (1999) Is this patient hypovolemic? *JAMA* 281(11): 1022-29

Mentes JC and Culp K (2003) Reducing hydration-linked events in nursing home residents *Clinical Nursing Research* 12(3): 210-25

Mentes JC, Wakefield BJ, Digglemann L and Culp K (2001) Assessing hydration status in older veterans and frail nursing home residents *J Am Geriatric Society* 49(4): S68

Mentes JC (2000) Hydration management protocol *Journal of Gerontological Nursing* 26 (10), pp. 6-15.

Molaschi M, Ponzetto M, Massaia M, Villa L, Scarafiotta C and Ferrario E (1997) Hypernatremic dehydration in the elderly on admission to hospital *The Journal of Nutrition Health and Aging* 1(3): 156-60

Moriarty D and Huson E (2001) Hypodermoclysis for rehydration in the community *British Journal of Community Nursing* 6(9): 437-45

Mukamel M (1997) Risk-adjusted outcome measures and quality of care in nursing homes *Medical Care* 35(4): 367-85

Mion LC and O'Connell A (2003) Parental hydration and nutrition in the geriatric patient: clinical and ethical issues *Journal of Infusion Nursing* 144-52: 144-52

Noble-Adams R (1995) Dehydration: subcutaneous fluid administration *British Journal of Nursing* 4(9): 488-94

Rafferty M (2004) *Dehydration: A practice guide for nurses* Glasgow: Greater Glasgow Health Board

Robinson SB and Rosher Rb (2002) Can a beverage cart help improve hydration? *Geriatric Nursing* 23(4): 208-11

Rochon PA, Gill SS, Litner J, Fischbach M, Goodison AJ and Gordon M (1997) A systematic review of the evidence for hyperdermoclysis to treat dehydration in older people *Journal of Gerontology Series* A Biological Sciences and Medical Sciences 52A (3): M169-M176

Sandberg O, Gustafson , Brannstromm B et al (1999) Clinical profile of delirium in older patients *JAGS* 47: 1300-06

Sasson M and Shvartzman (2001) Hyperdermoclysis: An alternative infusion technique *American Family Physician* 64(9): 1575-78

Simmons SF, Alessi C and Schnelle JF (2001) An intervention to increase fluid intake in nursing home residents: prompting and preference compliance *JAGS* 49: 926-33

Slesak G, Schurle JW, Kinzel E, Jakob J and Dietz PK (2003) Comparison of subcutaneous and intravenous rehydration in geriatric patients: a randomised trial *JAGS* 51(2): 155-60

Steiner N (1998) Models of hydration in palliative care patients *J Palliat Care* 14: 6-13

Suhayda R and Walton JC (2002) Preventing and managing dehydration *MedSurg Nursing* 11(6): 267-78

Thomas M and Day R (2003): A good death: quality of death can be measured outside hospices *BMJ* 320: 7243: 1205

Thomas DR, Tariq SH, Makhdomm S, Haddad R and Moinuddin A (2004) Physicians misdiagnosis of dehydration in older adults *Journal of American Directors Association* 4(5): 251-4

Tolson D, Smith Mand Knight P(1999) An investigation of the componenets of best nursing practice in the care of acutely ill hospitalized older patients with coincidental dementia: a multi-method design *Journal of Advanced Nursing* 30(5): 1127-36

Zeleznik J (2001) Delirium: still searching for risk factors and effective preventive measures? JAMA 49(12): 1729-32

Further Reading

Adunsky A, Levy R, Heim M, Mizrani E and Arad M(2003) The unfavourable nature of preoperative delirium in elderly hip fractured patients Archives of gerontology and Geriatrics 36(1):67-74

Alberta Palliative CareAdministering hyperdermoclysis http://www.Albertapalliative.net/apn/pchb/k_AdministeringHypodermoclysis.html

Armstrong–Esther CA, Browne KD, Armstrong Esther DC and Sander L(1996) The institutionalised elderly: dry to the bone *International Journal of Nursing Studies* 33: 619-28

Armstrong L, Maresh C, Castellanti J Bereron M Kenefick R La Gasse K et al (1994) Urinary indices of hydration status *International Journal of Sports Nutrition* 4:265-279

Black S (2002) Safe Home Nursing Standard 16(25): 16-17

Black D and Pearson M (2002) Average length of stay, delayed discharge and hospital congestion BMJ 325(7365) :610-12

Brown M (2000) Hypodermoclysis : another way to replace fluids *Nursing* 30(5): 58-9

Cambell Taylor I () Dehydration in long-term care :causes and a solution www.seniorhealthcare.org

Coleman EA, Barbaccia JC and Croughan- Minihane MS(1990) Hospitalization rates in nursing home residents with dementia JAGS 38:108-12

Copeman J (1999) *Nutritional care for older people a guide to good practice* London: Age Concern

Dasgupta M Binns MA, Rochon PA (2000) Subcutaneous infusion ina long term care setting *JAGS* 48:795-9

Donnelly M (1999) The benefits of hypodermoclysis *Nursing Standard* 13(52:15-21

De Conno F, Panzeri C, Brunelli C, Saita L and Ripamonti C (2003) Palliative care in a national cancer center: results in 1987vs 2000 *Journal of Pain and Symptom Control* 25(6):499-511

Douglas C (1999) Bedblocking for beginners BMJ 318(7200): 1773

Elder A (2001) Can we manage more acutely ill elderly patients in the community? *Age and Ageing* 30: 441-3

Evans D, Hodgekinson B and Wood J(2003) Maintaining oral hydration in older adults: a systematic review *International Journal of Nursing Practice 9(3)June 2003 S19-S28*

Fainsinger RL, MacEachern T, Miller MJ, Brurera E, Spachynski K, Kuen N and Hanson (1994) The use of hypodermoclysis for rehydration in terminally ill cancer patients *Journal of Pain and Symptom Control* 9(5): 298-302

Scottish Executive (2001) *Fair Care for Older People Care Development Group Report September 2001* Edinburgh: Scottish Executive: 20

Farley PC, Kam-Lung-L and Suba S (1986) Severe Hypernatremia in a patient with psychiatric illness *Arch Int Med* 146:1214-1215

http://familydoctor.org/ Artificial hydration: Hyperdermoclysis What is hyperdermoclysis?

Feinsod Fm, Levenson SA, Rapp K Rapp MP, Beechinor E and Liemann L (2004) Dehydration in frail older residents in long term care facilities *Journal of American Medical Directors Association* 5(2) :S35-41 Mar April

Francis J (1997) Outcomes of Delirium: Can systems of care make a difference? JAMA 45(2):247-48

Gasper P (1999) Water intake in nursing home residents *Journal of Gerontological Nursing* 25(4): 22-9

Gross CR, Lindquist RD, Woolle AC et al (1992) Clinical indicators of dehydration severity in elderly patients *J of Emergency Medicine 10: 267-74*

Goodwin DD, Hanson JC and Berry CP (2002) The changing face of Canadian home parental therapy *Journal of Infusion Nursing* 25(6):372-8

Grimley-EvansJ and Tallis RC (2001) A new beginning for care of elderly people BMJ 322(7290) :807-9

Innouye SK, Foreman MD, Mion LC, Katz KH and Cooney LM (2001) Nurses' recognition of delirium and its symptoms Archives of Internal Medicine 161(20): 2467-73

Innouye SK, Bogardus ST, Charpentier PA, Leo-Summers Let al (1999) A multicomponent intervention to prevent delirium in hospitalized older patients The New England Journal of Medicine Boston: 340(9): 669-78

Jiwa M, Gerrish K Gibson A and Scott H (2002) Preventing avoidable admission of older people British Journal of Community Nursing 7(8) :426-31

Jester R and Turner D (1998) Hospital at home : the Broomsgrove experience Nursing Standard 12(20): 40-2

Jordon S and Torrance c(1995) Bionursing :confusion in elderly people *Nursing Standard* 10(6): 30-2

Kayley Jand Findlay (2003) Vascular access devices used for patients in the community *Community Practitioner* 76(6) :228-31

Kayser –Jones J, Schell E, Porter C, Barbaccia J and Shaw H (1999) Factors contributing to dehydration in nursing homes: inadequate staffing and lack of professional supervision *Journal of Advanced Nursing* 47:1-14

Koffman J, Fulop N and Hudson M(1996) Assessing the impact of delayed discharge on acute care Nursing Standard 10(20):41-3

MacKay K (2002) *Helping Housebound Older People to Eat: Literature Review* (Internal Document) Stirling: Department of Applied Social Science

Macklin D (2003) Phlebitis : a painful complication of peripheral IV catheterization that may be prevented *American Journal of Nursing* 103(2):55-60

Mansfield S, MonaghanH and Hall J (1998) Subcutaneous fluid administration and site maintainence *Nursing Standard* 13(12):56-62

McDonald F (2004) Liatrogenic prescribing in acute care - learning from our mistakes *Journal of Gerontological Nursing* 30(3): 20-25

Mentes JC, Culp K, Mass M et al (1999) Acute confusion indicators: risk factors and prevalence using MDS data *Res Nurs Health* 22: 85-105

Musson ND, Kincaid J , Ryan P et al (1990) Nature, Nuture nutrition : interdisciplinary programa to address the prevention of malnutrition and dehydration *Dysphagia* 5:96-101

Mitty E and Mezey M 91998) Integrating advanced practice nurses in home care:recommendations for a teaching program *Nursing and Health Care Perspectives* 19(6):264-70

Moriarty D and Hudson E (2001) Hyperdermomoclysis for rehydration in the community *British Journal of Community Nursing* 6(9) 437-45

Muck E, Lyle RM, Galer-Unti R(2004) Knowledge and Behavior change in non-institutionalized elderly following and educational intervention on hydration Pedagogy and Special Population Posters The Preliminary Program for 2004 AAHPERD National Convention and Exposition http:/ /aahperd.confex.com/aahperd/preliminary program/abstract

Nazaro L 92000) How age affects fluid intake Nursing Times 96(31: 8

Nobel-Adams (1995) Dehydration: subcutaneous fluid administration *British Journal of Nursing* 4(9): 488-94

O'Keefe ST, and Lavan JN(1996) Subcutaneous fluids in elderly hospital patients with cognitive impairment *Gerontology* 42:36-39

Palevsky PM, Bhagrath R and Greenberg A (1996) Hypernatraemia in hospitalised patients *Ann Internal Medicine* 124:197-203

Potkins D, Bradley S, Shrimanker J, O'Brien J, Swann Aand Ballard C(2000) End of life treatment decisions in people with dementia:carer's views and the factors which influence them International Journal of Geriatric Psychiatry 15: 1005-8

Proctor S, Wilcockson J, Pearson P and Allgar v(2001) Going home from hospital: the carer/patient dyad Journal of Advanced Nursing 35(2):206-17

Seymour DG, Henschke PJ, Cape RDT and Campbell AJ (1980) Acute confusional states and dementia in the elderly: the role of dehydration/ volume depletion physical illness and age *Age and Ageing* 9:137-146

Shepperd S and Illiffe S (2004) Hospital at home versus in-patient hospital care (Cochrane Review) The Cochrane Library Issue 2

Simmons SF and Reuben D (2000) Nutritional intake monitoring for nursing home residents: a comparison of staff documentation, direct observation and photographic methods *JAGS 48:209- 13*

Sudeep G, Dasgupta M and Rochon P (2001) Hypodermoclysis in the treatment of hydration *Americal Family Physician* www.aafp.org

The AM, Pasman R , Onwuteaka- Philipson B, Ribbe M and Van der Wal (2002) Withholdingthe artificial administration of fluids and food from elderly patients ewith dementia :ethnographic study *BMJ* 325(7376) :1326

Terol E, Hamby E , Sacristan A and Sanz F et al (2001) An organizational model for a primary care program of home health care in Spain: a description of the program The Health Care Manager 20(2): 18-27

Urie J, Fielding H, McArthur D, Kinnear M, Hudson S and Fallon M (2000) Palliative care The Pharmaceutical Journal 265(7119): 603-14

Valios N (2000) How a rapid response team cabn keep beds free Community Care 31 Aug 28-9

Van de Steen J, Kruse RL, Ooms ME, Ribbe M and van der Wal (2004) Treatment of nursing home residents with dementia and lower respiratory tract infection (LRI) in Missouri and the Netherlands *JAGS* 52(5) :691-99

Vetter N (2003) Inappropriate delayed discharge from hospital: what do we know? BMJ 326(7395): 927-9

Victor C, Healy J, Thomas A and Seargeant J(2000) Older patients and delayed discharge from hospital Health and Social care in the Community 8(6):443-52

Waitt C, Waitt P and Pirmohamed M (2004) Intravenous therapy *Post Graduate Medical Journal* 80(939) :1-6

Weinberg AD, Pals JK, Leveresque PG et al (1994) Dehydration and death during febrile episodes in the nursing home *JAGS 42:968-71*

Wentworth J "003) Withholding artificial hydration at the end of life: clinical and ethical considerations for terminally ill patients University of Pennsylvania

Wilson A, Parker H, Wynn A and Jagger C et al (1999) Randomised controlled trial of effectiveness of Leicester hospital at home scheme compared with hospital care BMJ 319(7224) : 1542-46

Woodhall HE (2002) Alternatives to rehydration during hypodermoclysis *Am FAM Physician* 66(1): 1575-8

Woodrow P (2003) Assessing fluid balance in older people: fluid replacement *Nursing Older People* 14(10) :29-30

Yen P (2003) Maintaining cognitive function with diet (Preventing or slowing the advance of Alzeimer's disease, dementia and cognitive impairment *Geriatric Nursing* 24(1) 62-3

Zimmerman S, Gruber-Baldini AL, Hebel JR, Sloane PD and Magaziner J (2002) Nursing home facility risk factors for infection and hospitalization: importance of registered nurse turnover, administration and social factors JAMA 50(12):1987-95

Zwicker CD (2003) The elderly patient at risk *Journal of Infusion Nursing* 26(3): 137-43

Zembrzuski C (1997) A three dimensional approach to hydration of elders: administration, clinical staff and in-service education *Geriatric Nursing* 18:20-6